Nature's Habitats

IN THE RAINFOREST

Annabel Griffin

Illustrated by Rose Maclachlan

First published in Great Britain in 2024
by Hungry Tomato Ltd
F15, Old Bakery Studios,
Blewetts Wharf, Malpas Road
Truro, Cornwall, TR1 1QH, UK

Copyright © 2024 Hungry Tomato Ltd

No part of this publication may be reproduced, stored in a retrieval system, or transmitted in any form or by any means, electronic, mechanical, photocopying, recording, or otherwise, without prior written permission of the copyright owner.

A CIP catalogue record for this book is available from the British Library.

ISBN 9781835693537

Printed in China

Discover more at:
www.hungrytomato.com

Psst! A little frog is hiding on every page. Can you spot us all?

CONTENTS

In the Rainforest	4
Radiant Reptiles	6
Going Ape	8
Brilliant Birds	10
Plant Paradise	12
Hanging Around	14
Amazing Mammals	16
Life Amongst the Leaves	18
Where in the World?	20
Did You Know?	22
Who was Hiding?	23
Glossary	24

Words in bold capital letters **LIKE THIS** can be found in the glossary.

RADIANT REPTILES

Rainforests are warm and wet, which makes them the perfect HABITAT for lots of different REPTILES.

A sixth sense
These snakes have special sensors that allow them to 'see' heat. This helps them hunt live PREY.

Colour changers
Some chameleons can change colour. They do this for CAMOUFLAGE and to communicate with others.

Eyelash Viper

Chameleon

Tree huggers
These large snakes live high up in the trees and wrap themselves around branches.

Emerald Tree Boa

Sticking around
Geckos have lots of tiny hairs on the bottoms of their feet, which help them grip onto branches.

Gecko

Getting defensive
Iguanas are large lizards with powerful tails and sharp teeth, but they are actually harmless HERBIVORES.

Green Iguana

GOING APE

Apes are our closest relatives in the animal kingdom. Sadly, all of these apes are ENDANGERED because of habitat destruction and POACHING.

Going grey
The hair on a male gorilla's back turns grey as it gets older, giving them the name *silverback*.

Gorillas

Male Gorilla

Leading the troops
Gorillas are SOCIABLE and live in groups known as *troops*, led by the strongest male silverback.

Swinging in the trees
Orangutans have very long, strong arms, which are great for swinging from branch to branch.

Baby bonding
Baby orangutans stay with their mothers for up to 7 years.

Orangutans

We are family
We are most closely related to chimpanzees. They are very intelligent and can make and use tools.

Chimpanzee

BRILLIANT BIRDS

There are lots of colourful, exotic birds that fly among the tropical treetops.

Colourful couples
Macaw's are parrots that MATE for life. They form strong bonds and look after their young together.

Scarlet Macaw

Rhinoceros Hornbill

Blow your horn
This unusual horn is hollow and makes the bird's call much louder, a bit like a trumpet.

Powerful hunters
These massive BIRDS OF PREY have a WINGSPAN that can reach over 2 metres!

Harpy Eagle

Toucan

Toucy fruity
A toucan's large, colourful beak (or *bill*) is a useful tool for peeling fruit.

Whirring wings
A hummingbird's wings can beat about 70 times per second, making a humming noise.

Hummingbird

PLANT PARADISE

Rainforests are some of the greenest places on Earth, and are home to thousands of different trees and plants.

Twists and twirls
There are more than 2,500 different types of vine that grow in rainforests. They wrap around and hang from trees.

Don't worry, this frog is not being eaten! It is waiting for some tasty insects to come by.

Vines

Pitcher Plant

Meat-eating plant
Pitcher plants are CARNIVOROUS, which means they eat animals. They trap insects inside their bowl-like pitcher.

Banana Plant

Cocoa pod

Fruits of the forest?
Banana plants look like trees but they are actually giant herbs!

Cocoa Tree

A tasty tree
The pods on this cocoa tree are filled with beans that are used to make chocolate. Yum!

Orchids

Blooming beautiful
Rainforests contain over 10,000 different kinds of orchid.

HANGING AROUND

Do you like climbing trees? These animals spend lots of time hanging out in their branches.

Laidback life
Sloths are the slowest **MAMMALS** on the planet and spend most of their life hanging upside down.

Sloth

Squirrel Monkeys

Jungle piggyback
Squirrel monkeys carry their babies on their backs as they travel through the trees.

Loud mouth
Howler monkeys live high up in the trees and make deep, loud calls that can be heard up to a mile away!

A balancing act
These lemurs are excellent climbers and use their stripy tails to help them balance.

Head rush
Flying foxes sleep upside down with their wings wrapped around their bodies.

AMAZING MAMMALS

These large, four-legged mammals all make their homes on the forest floor, but tigers, jaguars, and anteaters can climb trees too!

Tiger print
Just like fingerprints, every tiger's stripes are unique.

Bengal Tiger

Tip-top tongue
Anteaters use their long tongues to slurp up about 35,000 ants and termites a day!

Giant Anteater

Jungle athletes
Jaguars are good at both climbing and swimming.

Why the long face?
A tapir's flexible trunk is perfect for stripping leaves off branches.

World's largest rodent
Capybaras look a bit like giant guinea pigs, and they are actually closely related.

LIFE AMONGST THE LEAVES

It could be easy to overlook some of the incredible creepy-crawlies hiding in the forest.

Elephant Beetle

Brilliant beetles
The male elephant beetle's horns are used as protection from **PREDATORS** and to compete with other males.

Praying Mantis

Praying or preying?
Praying mantises have sharp spikes on their front legs to help them catch and kill other insects.

Leaf choppers
These ants have large jaws for cutting up pieces of leaf to carry back home.

Blue Morpho Butterfly

Leafcutter Ants

Big, blue and beautiful
This stunning butterfly is one of the largest in the world, with a wingspan of 12-20 cm.

Goliath Birdeater

King of the spiders
This tarantula is the biggest spider on the planet! It is large enough to eat small birds, but it usually eats insects.

WHERE IN THE WORLD?

The Amazon Rainforest, in South America, is the biggest tropical rainforest in the world. Millions of plants and animals live there.

This map shows all of the rainforest areas in the world. Not all rainforests are the same. Different plants and animals are found in each one.

ASIA

AUSTRALIA

KEY

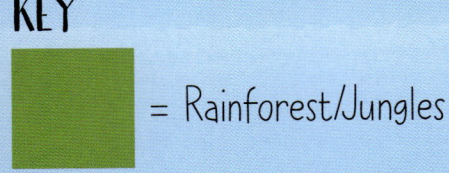 = Rainforest/Jungles

21

DID YOU KNOW?

Smell you later!
Male ring-tailed lemurs will have 'stink fights' by covering their tails in a smelly perfume from their wrists and then waving them at each other until one of them gives up and runs away!

Amazing abilities
Hummingbirds are the only birds in the world that can fly backwards and upside down!

A twist in the tail
Geckos and iguanas will snap their tails off if a predator grabs them. Don't worry, their tails grow back!

WHO WAS HIDING?

Did you spot the little poison dart frogs playing hide-and-seek in each rainforest scene?

Poison dart frogs can be found in rainforests in Central and South America.

They are tiny. Some are only the size of a human thumbnail!

They are one of the most poisonous animals in the world. Just one lick could be fatal!

They come in lots of different bright colours and patterns that act like a warning sign to possible predators.

Don't eat me! I'm very poisonous!

GLOSSARY

birds of prey - birds that mainly eat meat.

camouflage - to look like something else so as not to be easily seen.

carnivorous (carnivores) - animals (or plants!) that mainly eat meat.

endangered - if a type of animal or plant is in danger of dying out forever, then they are endangered.

habitat - where an animal or plant lives.

herbivores - animals that only eat plants.

mammals - animals with specific features. They all have hair or fur, drink milk from their mothers as babies, have a backbone, and are warm-blooded (they can keep their bodies warm, even when it's cold outside). Humans are mammals too!

mate - when animals form a pair that they have babies with.

poaching - illegal hunting of animals by humans.

predators - animals that hunt and kill other animals for food.

prey - an animal that is hunted by other animals for food.

reptiles - are animals with specific features. Reptiles have dry skin with scales, a backbone, breathe using lungs and are cold-blooded.

sociable - animals and people that are friendly and like spending time with others.

wingspan - the distance from the tip of one wing to the other, when they are fully stretched out.

The Author
Annabel Griffin is a writer and artist based in London, UK. Having worked as a bookseller for many years, she is now working in the children's publishing industry. Annabel's most recent publications include *Seasons* and *The Spectacular Lives of Sharks*.

The Illustrator
Rose Maclachlan is an illustrator based in Devon, who graduated from Falmouth University with a BA in Illustration. She likes to experiment with collage and texture to create her work and takes inspiration from her love of the outdoors and the beach.

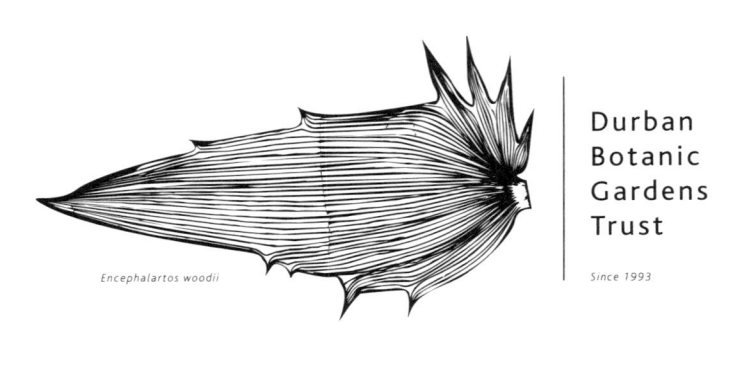

Encephalartos woodii

Durban Botanic Gardens Trust

Since 1993